I0756104

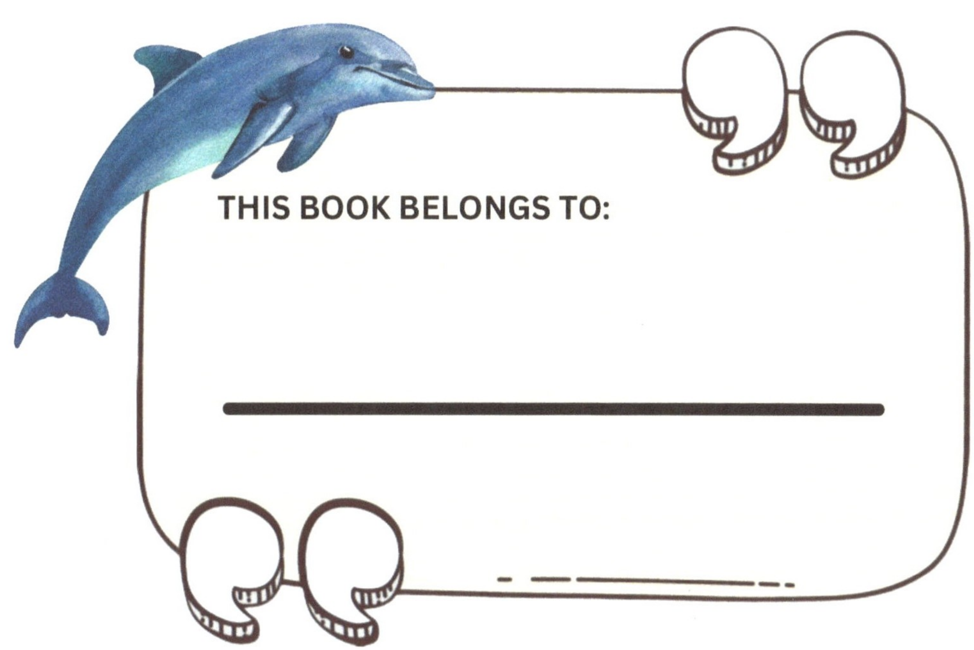
THIS BOOK BELONGS TO:

THE WONDERFUL
WORLD OF DOLPHINS
Mimi Jones

Dedicated to all the dolphin lovers.

ISBN 978-1-958985-92-2

www.joeysavestheday.com

A Mimi Book

Dolphins are mammals. That means they breathe air, give birth to live young, and nurse their babies with milk. Even though they live in the ocean, they share many traits with land animals.

Mammals

Breathe

They breathe through a blowhole. Located on top of their heads, this special opening lets dolphins take quick breaths when they surface. They can't breathe through their mouths like humans do.

Dolphins live in oceans and rivers. While most dolphins swim in saltwater seas, some species, like the Amazon river dolphin, live in freshwater. They adapt well to different environments.

Adapt means to make changes that help you stay safe and strong.

Their skin is smooth and rubbery. Dolphin skin feels like wet leather and helps them glide through water easily. It also heals quickly from scrapes and scratches.

Smooth

20 MPH

Dolphins are fast swimmers. They can reach speeds of up to 20 miles per hour. Their sleek bodies and strong tails help them zoom through the waves.

Intelligent

Dolphins are very smart. Scientists believe they are one of the most intelligent animals on Earth. They can solve problems, learn tricks, and even understand symbols.

Reflection

They recognize themselves in mirrors. This shows they have self-awareness, a trait shared with only a few animals. They know the reflection is not another dolphin.

Protect

Dolphins use tools. Some dolphins use sea sponges to protect their noses while searching the ocean floor. This clever behavior shows planning and creativity.

They play games. Dolphins enjoy playing with seaweed, bubbles, and even other animals. Play is a sign of intelligence and social bonding.

Games

Dolphins can learn hand signals. Trainers often teach dolphins to respond to gestures. This shows they can understand and remember human-made signs.

Trainer

Dolphins talk with sounds. They use clicks, whistles, and squeaks to communicate. Each sound has a different meaning, like calling a friend or warning others.

Clicks

Squeaks

Whistles

Greetings

Each dolphin has a signature whistle. This special sound acts like a name. Dolphins use it to identify and greet one another.

They use echolocation. By sending out sound waves and listening for echoes, dolphins can find fish and navigate dark waters. It’s like having built-in sonar.

Echolocation

Dolphins hear very well underwater. Their hearing is much better than humans, especially in the ocean. They can detect sounds from miles away.

Signals

Excitement

Tail slaps are signals. Dolphins sometimes slap the water with their tails to show excitement or danger. It's a loud way to get attention.

Danger

Dolphins eat fish and squid. Their diet depends on where they live and what's available. They are skilled hunters and often work together to catch food.

Bottlenose dolphins eat a lot. They consume 15 to 30 pounds of food each day. That's like eating dozens of fish every single day!

15 to 30 pounds

Safety

Baby dolphins are called calves. They are born tail-first to avoid drowning. Calves stay close to their mothers for safety and learning.

DOLPHINS

There are over 40 dolphin species. Each one has unique traits, like size, color, and habitat. Some live in oceans, others in rivers.

Orca

Amazon river

Striped

Humpback

Bottlenose dolphins are famous. They are often seen in aquariums and movies. Their friendly nature makes them popular with people.

FAMOUS

Amazon river dolphins are pink. Their color comes from blood vessels near the skin. They live in muddy rivers and have long snouts.

BEHAVIOR

Spinner dolphins twirl in the air. They leap and spin like acrobats. This playful behavior helps them communicate and bond.

LARGEST DOLPHIN SPECIES

Orcas are dolphins too. Also called killer whales, orcas are the largest dolphin species. They live in cold oceans and hunt in groups.

RESTING DOLPHIN

Dolphins sleep with one eye open. They rest half their brain at a time so they can keep swimming and stay alert.

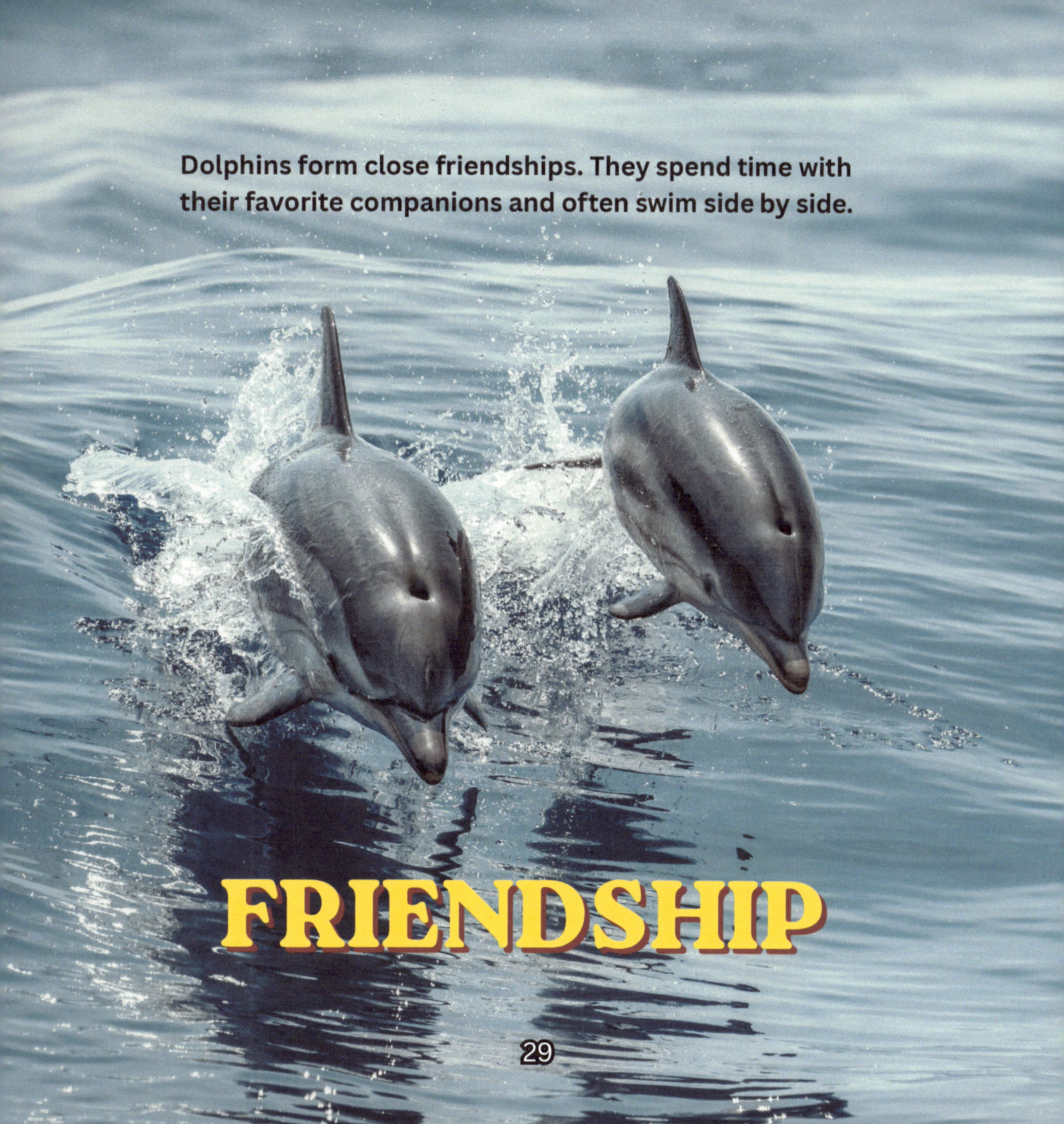

Dolphins form close friendships. They spend time with their favorite companions and often swim side by side.

FRIENDSHIP

Dolphins can leap 15 feet high. Their powerful tails launch them into the air. It's a spectacular sight!

15 FEET

ADJUST

They see well underwater. Dolphin eyes adjust to light and dark. They can spot fish and obstacles easily.

Dolphins don't smell. They lack olfactory nerves, so they rely on hearing and sight. Their sense of taste is limited too.

SENSES

They can hold their breath for up to 15 minutes. Dolphins surface often, but they can stay underwater for long periods when needed.

SURFACING

RESCUE

Dolphins help in rescue missions. Trained dolphins locate underwater objects and assist divers. Their skills are valuable to humans.

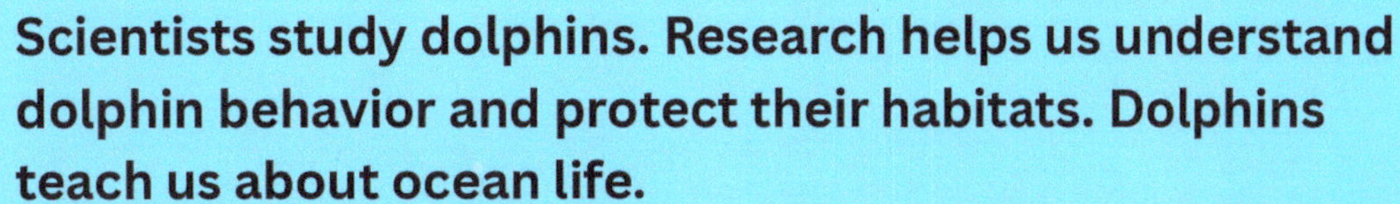
Scientists study dolphins. Research helps us understand dolphin behavior and protect their habitats. Dolphins teach us about ocean life.

OCEAN LIFE

Count the dolphins.

Thank you for exploring The Wonderful World of Dolphins with me. I hope you learned something new and had fun along the way.

If you enjoyed this book, please consider leaving a review. It helps other families discover it too.

See you in the next adventure!

Check out these other interesting books in the Wonderful World of series!

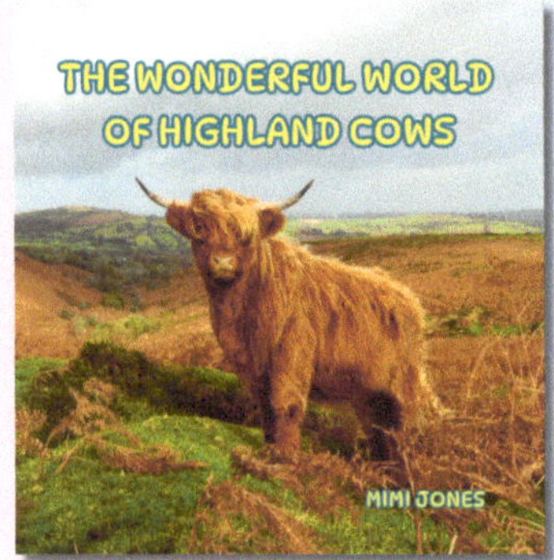

www.mimibooks.com

www.ingramcontent.com/pod-product-compliance
Lightning Source LLC
LaVergne TN
LVHW070157110826
845147LV00002B/435
9781958985922